Discovering
BEES AND WASPS

Christopher O'Toole

The Bookwright Press
New York

All photographs from Oxford Scientific Films

First Paperback Edition 1990
ISBN 0-531-18362-9

First published in the United States in 1986 by
The Bookwright Press
387 Park Avenue South
New York, NY 10016

First published in 1985 by
Wayland (Publishers) Limited
61 Western Road, Hove
East Sussex BN3 1JD,
England

© Copyright 1985 Wayland (Publishers) Limited

Library of Congress Catalog Card Number: 85-72247

Typeset by Planagraphic Typesetters Limited
Printed in Italy by G. Canale & C.S.p.A., Turin

Contents

Introducing Bees and Wasps
Bees, Wasps, and Their Relatives *8*
Looking at Bees and Wasps
    in Close-up *10*
The Nest *12*
Finding a Mate *14*

**Living Alone — Solitary Wasps
    and Bees**
Thread-waisted Wasps *16*
Mining Bees *20*
Mason Bees *22*
Leaf-cutter and Wool-hanger Bees *24*
Parasites and Galls *26*

**Living Together — Social Wasps
    and Bees**
What it Means to be Social *28*

Social Wasps and Hornets *30*
Bumblebees *32*
Honeybees — How the Colony
    Works *34*
Honeybees — A Worker's Life *36*

Enemies of Bees and Wasps
Cuckoo Bees and Wasps *38*
Other Enemies *40*

Bees, Flowers and People
Flowers and Bees are Partners *42*
Bees and People *43*

Glossary *45*
Finding Out More *46*
Index *46*

Introducing Bees and Wasps

Bees, Wasps and Their Relatives

Bees and wasps are common all over the world. The black and yellow wasps which sometimes spoil picnics are **social** insects. This means that they live together in groups called **colonies.** Honeybees and the fat, furry bumblebees are social too. So are ants, which are a special group of wasps.

But almost all of the 100,000 kinds of bees and wasps are **solitary.** Each nest is the work of a single female living alone.

Cuckoo bees, however, do not have nests, neither do all wasps. There are thousands of kinds of wasps that have no home. Some are called sawflies, and their grubs look very much like caterpillars.

Adult sawflies are usually rather dumpy insects and you can see them feeding on flowers. They are called sawflies because the female has a special kind of egg-laying tube, or **ovipositor,** which has teeth just like a saw. She uses this to make a slit in a leaf or stem into which she lays an egg.

Many thousands of wasps have grubs that feed inside or on the bodies of

A common wasp drinking nectar from ivy flowers.

other insects and spiders. They are called parasitic wasps. The adults are much thinner than sawflies. The female uses her ovipositor to inject eggs into or onto the **host's** body.

Right *An adult sawfly, one of the many kinds of wasps.*

Below *Sawfly grubs live the same kind of life as caterpillars, chewing leaves.*

Looking at Bees and Wasps in Close-up

If we look closely at a bee or wasp, we can see that its body has three main parts, the head, **thorax** and **abdomen.** The whole body is enclosed in a kind of armor plating. This is the skeleton. Unlike ours, the skeleton is on the outside of an insect's body. It is made of a light but strong horny substance called chitin.

Below *You can see the hairs and a fully-laden pollen basket on this honeybee worker.*

Above *A bumblebee's sting with a drop of poison on the end of it.*

Like all insects, bees and wasps have six legs. Their eggs hatch into grubs, which change into **pupae** before becoming adults. Bees and wasps also have two pairs of wings.

Each female of the nesting bees and wasps has a sting at the tip of the abdomen. It is really an ovipositor which has lost its egg-laying job. Instead, it is used in defense to inject a painful poison into enemies, or to paralyze prey.

All nesting wasps are hunters, feeding their grubs on insect **prey.** Bees are really a special group of hunting wasps that have given up catching insects. Instead, they feed their young on **pollen** and **nectar** collected from flowers.

Bees are much furrier than wasps. They are covered with branched hairs that trap pollen grains between them when the bee walks over flowers. Each bee uses her feet to scrape up the pollen grains and pack them into special pollen baskets which are either on her hind legs or under her abdomen.

The Body of a Honeybee

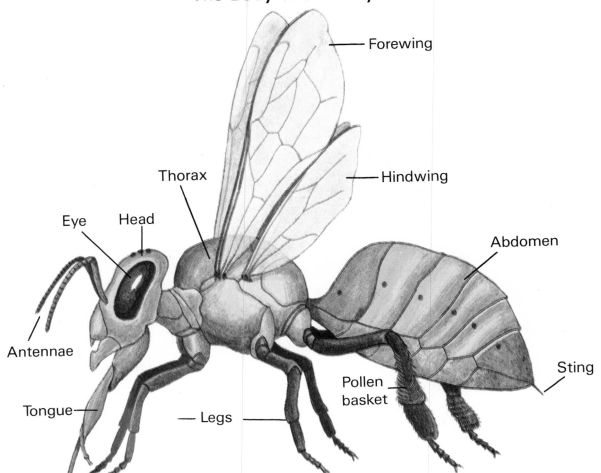

Above *Sand bees, which are solitary bees, dig holes in the soil to make their nests.*

Common social wasps make nests out of tiny pieces of chewed wood.

The Nest

The nest is a safe place built by bees and wasps. Here they store food for their young and lay their eggs. In cool countries, when the young have become adults, they may stay in the nest over winter. Others winter over as fully grown grubs. In tropical countries, an insect's life cycle is not usually interrupted by changes in temperature.

There are many different kinds of nests. Some bees and wasps dig burrows in the soil while others use holes in wood, made by wood-boring insects. These hole-nesters sometimes use man-made sites like garden hoses or crevices between window frames. Other bees and wasps make nests out in the open, on leaves, twigs or even in houses. They often use mud as a building material.

Any animal with a nest has to be able to find its way home. Bees and wasps do this the same way people do. Imagine you are going home from school. You

know the way without having to think about it. This is because you have remembered clues called landmarks. It may be something like this: turn left at the school door and left at the corner, then right at the post office.

When a female wasp or bee has finished her nest, she makes a special flight

The mud-dauber wasp makes her nest out of damp mud collected from the edges of pools.

around the entrance and remembers all the local landmarks. These may be things like trees, rocks and hills. She also uses the position of the sun in the sky to help her find her way home.

Finding a Mate

Solitary bees and wasps mate soon after they come out of their nests. A male does not help in nest building or finding food, so finding a mate is the only job he has to do. Male solitary bees and wasps mate many times. However, a male honeybee mates only once. After mating the male dies. Females usually mate only once; they store all the sperm they need in a little bag inside the body.

There are several different ways that male wasps and bees try to find mates. The males of some kinds simply fly among flowers looking for females.

Male solitary wasps usually emerge before the females. They often gather where an unmated female is digging her way out.

Other males spend most of their time around the nest sites waiting for females to emerge.

Bumblebee males fly around special routes called circuits. They stop regularly to put a scent on a leaf or stem from **glands** in their heads. The scent attracts unmated females who mate with the next male that comes flying around the circuit.

Male wool-hanger, or carder bees find

A male wool-hanger bee waits on a lamb's ears leaf for a female to enter his territory.

a place that has lots of flowers that the females like. Each male has a special place, called a territory, among the flowers. He chases away other males and even other insects that might want to visit the flowers. A male carder bee mates with any female that enters his territory.

Living Alone - Solitary Wasps and Bees

Thread-waisted Wasps

There are many thousands of kinds of solitary wasps and bees that dig out nests in the soil. Although you may find hundreds nesting in the same patch of bare earth, each nest is the work of a single female working alone.

One common type of hunting wasp that does this is called *Ammophila*, (Am-off-illa). The name means "sand lover" because most of these wasps like to nest in dry, sandy places.

Ammophila wasps are very long and thin, with narrow, thread-like waists, which is why they are called thread-waisted wasps. A thread-waisted wasp's day begins just after sunrise. She will have spent the night hanging by her jaws from a grass stem. Hundreds of thread-waisted wasps, both male and female, sleep together in big groups.

Several thread-waisted wasps waking up as the sun rises.

When a female thread-waisted wasp finds a good place for her nest, she digs a short tunnel. At the bottom she clears out a chamber called a **cell**. Here she will store caterpillars for her young to eat. It takes her a whole day to dig her nest, so when she has finished, it is time to fly back to the roost.

Below *A thread-waisted wasp closing her nest with a pellet of earth.*

Above *A female thread-waisted wasp digging a tunnel.*

The wasps soon warm up in the sun and start to fly. The males look for unmated females. The roosting females have already mated. They fly off to find a place to dig a nest.

Above *A female thread-waisted wasp with a caterpillar for her nest.*

The next morning, the thread-waisted wasps are again awakened by the sun's warmth. The females fly off to inspect their nests. They know the way because they remember all the landmarks from the day before. They need energy, so they visit flowers regularly during the day to feed on nectar.

After making her nest, a female thread-waisted wasp hunts for caterpillars among leaves and grass stems. She is very good at finding them. As soon as she has found one, she stings it in several places. This paralyzes it so that it cannot move. She flies back with it to her nest.

If the caterpillar is very heavy, she drags it back to her nest along the ground.

Back at the nest, she puts the caterpillar in the cell and lays an egg on it. Then, she catches several more, one at a time, and adds them to the first. Sometimes, other wasps try to steal a caterpillar and there is quite a tug-of-war.

When she has enough caterpillars, she closes the nest with soil and small pebbles.

Above *A section through the nest of a thread-waisted wasp shows several caterpillars, one of which has had an egg laid on it.*

Below *Female thread-waisted wasps stealing another female's caterpillar.*

Then she flies off to look for a suitable place to start another nest. She dies before her young are fully grown.

Soon, the egg hatches and the wasp grub eats the caterpillars. When it has finished it spins a **cocoon** and pupates. During this time the grub is broken down and reformed into an adult. A new wasp will emerge the following summer.

Mining Bees

Spring and early summer are the best times to see solitary mining bees. There are thousands of different kinds. Some come out in very early spring and you can find them visiting pussy willow blossoms. Many types are very choosy and visit only one kind of flower.

Like thread-waisted wasps, mining bees nest in the ground and make cells at the ends of the tunnels. They collect pollen and nectar to store in the cells for their young to feed on. They store the cells one at a time and lay an egg in each. A finished nest may have six to eight cells.

A female sand bee digs out her nest.

Hundreds of mining bees may nest close together in a patch of ground. One common kind is called the tawny mining bee. Large numbers often nest in garden lawns. The female is covered with a lovely coat of fox-colored red hairs. The males are thinner and have brown hairs. They fly around the nesting places looking for mates.

This female tawny mining bee has a coat of red hairs.

Other kinds of mining bees are not so hairy. The sand bee has a striped abdomen. The stripes are made of very short, dense, velvety hairs. You can find sand bees in summer. They nest in sandy places by the ocean or in old sandpits.

Mason Bees

Mason bees are very different from mining bees. Instead of digging nests in the ground, they find ready-made holes.

A female mason bee collecting mud, which she will use to close each cell of her nest.

These may be in hollow stems or beetle borings in dead wood. Sometimes, they nest in old walls. A few kinds are very choosy and nest only in empty snail shells.

The red mason bee is common everywhere. Like all mason bees, she has her pollen basket under her abdomen rather than on her hind legs.

A mason bee grub in its cell, together with its food, a mixture of pollen and nectar.

The nest of the red mason bee is very interesting. She makes a row of cells, one at a time, starting at the back of the tunnel. She fills the first cell with pollen and a little nectar. Then she lays an egg on the food and closes the cell with a little mud she has collected. She repeats this until she has filled the tunnel with a row of perhaps eight or ten cells. She closes the nest with a plug of mud.

When the grubs have eaten all the food, they spin a silk cocoon and pupate. When they are ready, the new adults bite their way out of the cells. In a cold country the new adults spend the winter in their cells and bite their way out the following spring.

Not all mason bees use mud. Some use a paste of chewed leaves or petals. Others use **resin** and many build nests in the open, like mud-dauber wasps.

Leaf-cutter and Wool-hanger Bees

You may sometimes have seen leaves with round and oblong holes cut out of their sides. This is probably the work of female leaf-cutter bees. They use the leaf pieces to line their nest tunnels. Like mason bees, the leaf-cutter nests in ready-made holes and has the pollen basket on the underside of its abdomen.

The rose leaf-cutter bee is about a centimeter (half an inch) long and is black, although it sometimes looks very yellow when it is dusted with pollen. The female cuts an oblong piece of leaf and flies back to the nest with it. She cuts several pieces until she has enough to make the side walls of her cell.

Two cells in the nest tunnel of a leaf-cutter bee, each with a young grub feeding on the store of pollen and nectar.

A female leaf-cutter bee cuts off a piece of leaf to make the cells of her nest.

A female wool-hanger, or carder bee, collecting hairs from a lamb's ears plant.

The bee nibbles the edges of the leaf to squeeze out some sap. She uses this as a kind of glue to stick the leaves together. Then she fills the finished cell with a runny mixture of pollen and nectar and lays an egg on the food. She plugs the cell with several round leaf pieces pressed closely together.

Leaf-cutter bees are found all over the world and there are many hundreds of kinds.

Wool-hanger or carder bees also visit leaves for nest-building materials. They use their jaws to scrape the hairs off hairy plants like lamb's ears. They use this "wool" to line their nests.

Parasites and Galls

There are many kinds of wasps that do not make nests. Instead, they let a plant do all the work. The female wasp lays some eggs in a leaf bud or in a stem. This makes that part of the plant grow in a

Below *A rose pincushion gall on a rose bush.*

Above *The inside of a gall, showing wasp grubs and pupae.*

very strange way. It grows all around the wasp eggs and becomes something called a gall. The wasp grubs live inside the gall and eat it.

One of the biggest mysteries in nature is how the wasp makes the plant grow a gall. We just do not know. Each kind of wasp causes a different kind of gall. One of the commonest is called a rose pincushion or moss gall. You can find it

Above *A cabbage white caterpillar being eaten by wasp grubs. The fluffy white objects are wasp cocoons.*

growing on wild roses.

Thousands of kinds of wasps have no nests at all. The females lay their eggs in or on the bodies of other insects. They are called parasitic wasps. The eggs hatch and the grubs feed on the body of the victim and eventually kill it.

Female parasitic wasps, such as the ichneumon (ick-new-mon), often have very long ovipositors. This means they can reach a host deep inside a log. Some kinds of parasitic wasps are useful to man

An ichneumon wasp bores through wood with her ovipositor to lay an egg on a wood-wasp grub.

because they attack pests like the caterpillars of the cabbage white butterfly, and the grubs of wood wasps, which destroy pine trees.

Living Together - Social Wasps and Bees

What it Means to be Social

Social bees and wasps live in family groups called colonies. Each colony has a single, egg-laying female called a queen and hundreds or even thousands of workers. The workers are female, too, but they do not lay eggs. They have many other jobs to do: they collect the food, build the nest and look after the young. Social bees and wasps do not seal their grubs up in cells with all the food they need. Instead, the young are fed every day by the workers.

Some of the smallest and simplest colonies are made by the slender paper wasps, called *Polistes* (Poll-ist-ees). There are hundreds of kinds and most of them live in the warmer parts of the world. The paper nest is started by the queen. She makes the paper by biting off and chewing little strips of dead wood and

A paper wasp queen and her workers on their nest under the eaves of a house.

A paper wasp visits the flower of a poinsettia plant for nectar.

mixing it with her **saliva.** The nest hangs by a little stalk from under a leaf or branch. It is made up of six-sided cells. The queen lays an egg in each cell and when the grubs have hatched, she feeds them from time to time on chewed up pieces of insects, usually caterpillars.

The first wasps to emerge are all females. They become workers and take over all the jobs the queen was doing, except egg-laying. The queen now spends all her time laying eggs and never leaves the nest. Later in the year, males and new queens are produced. They mate and the young queens leave to start new nests.

Social Wasps and Hornets

The black and yellow wasps that are attracted to sweet things like jam and rotting fruit are a type of paper-making wasp. They are sometimes called yellow jackets. Most kinds of social wasps are tropical, but yellow jackets are found mainly in cold countries. Yellow jackets make the same kinds of six-sided cells as *Polistes*. The cells are in layers called combs and the yellow jackets cover them with

This queen wasp has started to make a new nest on the outside of a house.

Above *A hornet having a drink of water.*

an outer envelope of paper. There is an entrance hole in the bottom of the nest. Some kinds build their nests in the open; others in holes underground. Sometimes they nest inside the roofs of houses and are a nuisance.

Hornets are very big yellow jackets. They like to nest in hollow trees and are not nearly as fierce as people think.

A yellow jacket's nest is started in spring by a mated queen that has **hibernated** over the winter. She behaves like a solitary wasp for the first few weeks until her first young emerge as adult wasps. They are workers. Unlike the

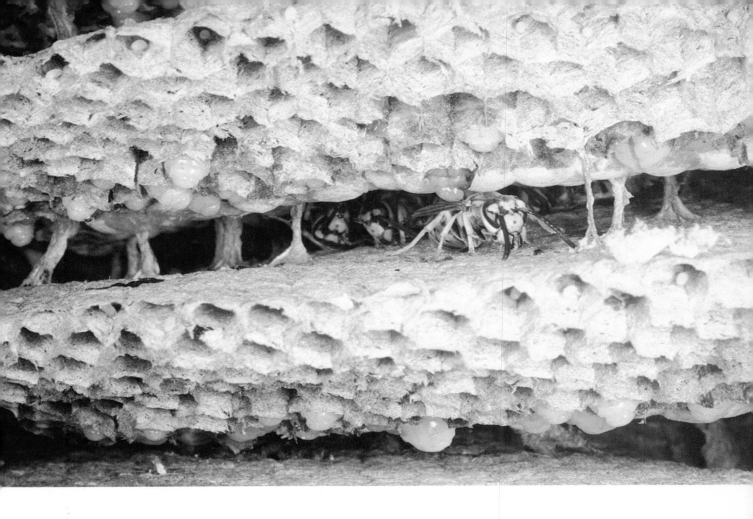

The inside of a common wasp's nest. Each layer consists of many cells, which contain wasps in various stages of development.

workers of the slender paper wasps, yellow jacket workers are smaller than the queen. A yellow jacket colony may have 15,000 workers.

The colony dies out at the end of the summer. Only the newly-mated young queens hibernate. They will each start a new colony the following spring.

Bumblebees

There are just over 200 kinds of bumblebees in the world. Most live in northern countries and you can see them from spring to summer wherever there are flowers.

Only newly-mated queens hibernate over the winter. In spring, you can see them bumbling around clumps of bushes and earth banks, looking for places in which to nest. Some types use old mouse nests; others make nests aboveground, in clumps of grass.

At first, the queen behaves like a solitary bee. She teases out a space in the dried grass of an old mouse nest and makes two little wax pots. She fills one with eggs and pollen mixed with a little honey. Then, she closes it with a sheet of wax. This is called a brood cell. The wax comes from glands in her abdomen. She stores honey in the second pot. Then she sits astride the brood cell and keeps the

Below *A bumblebee queen keeping her eggs warm. Her honey pot is next to her.*

Below *A bumblebee nest with cocoons and newly emerged adults.*

Above *A short-tongued bumblebee cuts a hole in a flower to steal the nectar.*

growing grubs warm. The honey pot is within easy reach and she does not have to leave the nest very often.

When the first workers emerge they take over food gathering and help the queen build more brood cells. Old brood cells are used to store pollen and honey. Eventually, the queen loses her ability to fly. She remains in the nest for the rest of her life, laying eggs.

In the summer, males and young queens are produced. The new queens mate and find a place to hibernate. The rest of the colony is killed off by the first frosts of autumn.

Honeybees — How the Colony Works

The large honeybee marked with a yellow dot is the queen. She is laying eggs.

Wild honeybees build nests in holes in rocks or in hollow trees. Beekeepers keep bees in special wooden nests called hives. A healthy colony contains one queen, 20,000-50,000 workers and up to 300 males called drones.

The nest is made of double-sided wax combs. Each comb contains hundreds of six-sided cells. The bees use the cells to store food and rear young.

The queen honeybee is larger than the workers and her only job is to lay eggs —

The egg of a honeybee in its cell. The queen lays one egg in each cell.

sometimes as many as 1,500 a day! The queen gives off a chemical called "queen substance," like the queens of yellow jacket wasps. This keeps the workers from making eggs, and from making special queen cells and rearing queens.

Whether a female grub becomes a worker or a queen depends on the food it is given. If it is to be a worker, it is fed for the first three days on royal jelly. This is made by glands in the heads of adult workers. After three days, the workers stop giving the grub royal jelly. Now it only eats pollen and honey. If a grub is to become a queen it is fed royal jelly throughout its growing period.

The workers only rear new queens if the old one begins to lay fewer eggs and does not make enough queen substance. When this happens, the old queen and some of the workers leave the hive in a swarm and start a new colony.

Meanwhile, the new queen mates with one or more drones on a special mating flight. Then she returns to the hive to begin a new life.

In northern countries, when winter comes, all the bees huddle together to keep warm, and feed on stored honey.

Fully-grown honeybee grubs in their cells. They are fed each day by worker bees.

Honeybees — A Worker's Life

The jobs a worker honeybee does depend on her age. First, she is a cleaner, then a nurse looking after the young. She then becomes a builder and later collects pollen and nectar from bees returning to the nest.

Next, she is a guard at the nest entrance. She attacks any enemies trying to get in and may die doing this. Her sting has little teeth which catch in the skin. When the guard bee struggles to free herself the sting stays behind in the enemy, and keeps on pumping poison. It also gives off a special alarm scent that brings other guard bees to the place of danger.

After the guard duty, the worker becomes a field bee. She spends the rest of her life collecting pollen and nectar. Field bees have two kinds of dances which tell other bees where to find food.

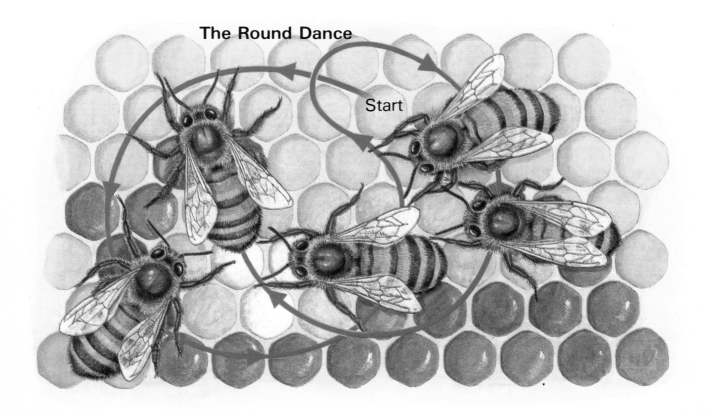

The Round Dance

Start

The round dance tells them that there are good flowers up to 90 meters (100 yards) away from the nest. A dancing worker runs around in small circles, often changing direction. The faster she changes direction, the better the food.

The wagging dance is for flowers that are more than 90 meters (100 yards) away. The bee dances a kind of flattened figure-eight. She wags her abdomen from side to side in the straight run between the two halves of the figure eight. The longer the wagging dance, the farther away are the flowers. And, the faster she wags, the better the food. The wagging dance also tells the other bees which way to go to find the flowers.

A worker bee drinking water from a pond.

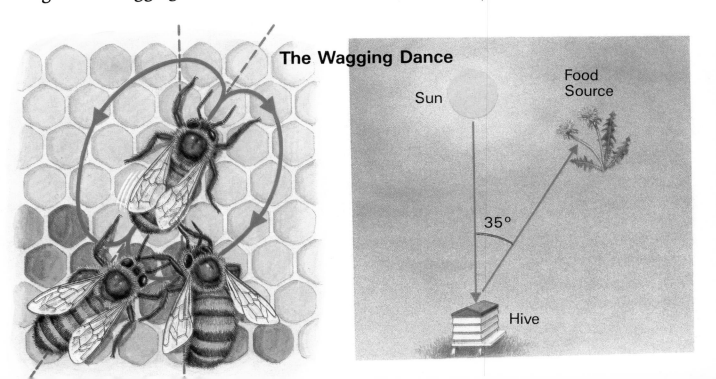

The Wagging Dance

Enemies of Bees and Wasps

Cuckoo Bees and Wasps

Cuckoo bees and wasps are cheaters, just like the cuckoo bird. Instead of making nests for themselves and looking after their young, they lay their eggs in the nests of other bees and wasps.

There are many different kinds of cuckoos and they are usually more brightly colored than their victims. They are very choosy. Each type has its favorite kind of wasp or bee. The female cuckoo waits by the nest entrance until the mother wasp or bee leaves to collect food. Then, she pops in and lays an egg. Afterward, she looks for other nests.

The cuckoo's egg hatches quickly.

Below *A cuckoo wasp searching for a host nest in which to lay its egg.*

Above *A female cuckoo bee has no pollen basket because she does not need to collect pollen to feed her young.*

The grub has very large jaws. It uses them to kill and eat the host egg or grub. Then it sheds its skin and loses its big jaws. Now it has tiny jaws like other wasp or bee grubs.

If the host is a wasp, the cuckoo wasp grub then eats the prey stored by the mother wasp. If the host is a bee, the cuckoo eats the stored pollen and honey.

Cuckoo bumblebees invade the nests of other bumblebees. They do this when the host queen has reared a lot of workers. The cuckoo bumblebee relies on the host workers to rear her young.

Other Enemies

Bees and wasps have many enemies. This is not surprising. A large nest has lots of juicy grubs to eat and a honeybee nest also contains a big store of honey. Chimpanzees and orang-utans raid honeybee nests. So do yellow jacket wasps, and you can sometimes see a toad sitting by a hive entrance catching bees as they come back to the hive.

Beehives are a favorite feeding place for the African honey badger or ratel. It has very strong claws for breaking open nests. It also has loose, rubbery skin, which seems to protect it from stings. Special stink glands give it an awful smell. Perhaps this helps to reduce the number of angry bees willing to attack it. All in all, the ratel is well-equipped to be an enemy of bees.

The African bee-eater is one of the few birds that eat bees. It knocks the bee against a branch to destroy the sting.

Sometimes the ratel is helped by a bird called the honey-guide. This bird needs to eat beeswax to survive. But it is not usually strong enough to break into a nest. When it finds a nest, it flies around until

The ratel, or honey badger, feeds on bee grubs as well as honey.

it comes across a ratel. Then, it makes a churring noise and does a little dance. This leads the ratel to follow the bird, which continues its churring and dancing until they both reach the nest. The ratel breaks open the nest or hive and has a meal, and the honey-guide can eat as well. Sometimes, the honey-guide attracts people to a nest and the same thing happens.

Bees, Flowers and People

Flowers and Bees are Partners

Flowers need bees to help them reproduce. Bees do this by carrying pollen from flower to flower. The pollen grains are made by the male parts of the flower. They fertilize eggs made by the flower's female parts. This is called pollination. Flowers make more pollen than they need for mating. The extra pollen is a food reward for the bees. They also make nectar as a reward. This is a sweet mixture of sugars, which gives the bees lots of energy; they make it into honey.

In South America, there are beautiful long-tongued bees, which pollinate orchids. Only the male bees do this. They pick up little packages of pollen when they crawl through the flower. The males are rewarded by a perfumed oil made by the orchid. The bees store the oil in their hind legs. Here, it is changed into a scent that attracts female orchid bees. The females find the males and the bees mate.

This bee collects pollen only from sunflowers.

A long-tongued orchid bee visiting an orchid.

Bees and People

People have been keeping honeybees for their honey for thousands of years. If you would like to keep bees, ask your teacher about starting a school beehive.

Farmers also use honeybees to pollinate crops. When fruit trees are in flower, farmers often pay beekeepers to put hives into their orchards. This

Hundreds of thousands of leaf-cutter bees will nest in this shed in an alfalfa field.

improves pollination and increases the amount of fruit produced.

Sometimes, farmers use leaf-cutter bees to pollinate alfalfa. This is an important food crop for cattle. The farmers put out blocks of wood drilled with holes for the bees to nest in.

Mason bees and leaf-cutter bees will nest in lengths of hollow garden cane, in large drinking straws packed into tin cans or in blocks of softwood drilled with holes. You can hang bundles of these "bee houses" on the sunny sides of garden sheds, garden walls or fence posts, especially near those places where bees may already be nesting. The bees will soon find them and it is fun to watch them coming and going. They will not sting unless they are handled roughly.

Above *You have to wear special clothing when you look after bees in a hive.*

Some "bee houses" you could make at home.

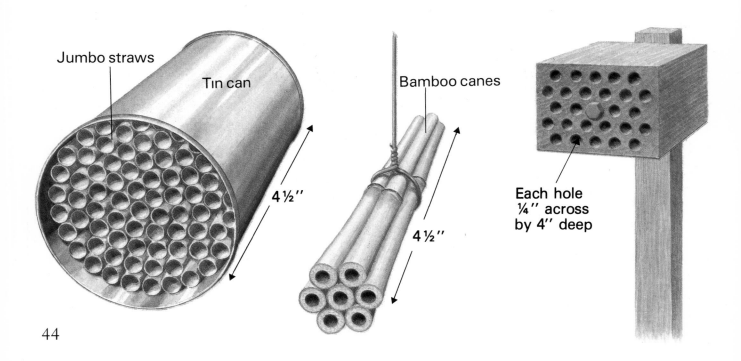

Glossary

Abdomen The third part of the three main body parts of insects (head, thorax and abdomen). It contains the stomach, sex organs, wax glands, sting and poison glands.

Cell A specially prepared space in the nest of a bee or wasp, where food is stored and an egg is laid and where the young bee or wasp grows.

Cocoon The silky case that many insect grubs make around themselves.

Colony The nest and family group of social insects.

Gland A special part of the body that makes a particular substance such as wax or scents.

Hibernate To sleep through the winter.

Host The victim of a parasitic wasp or cuckoo wasp or bee.

Nectar The sweet mixture of sugars provided by flowers to attract bees.

Ovipositor The egg laying tube of female insects. In some wasps and bees it has lost its egg-laying job and is a sting.

Pollen The dusty powder made by the male parts of flowers, containing the male sex cells.

Prey Creatures hunted by others for food.

Pupae Plural of pupa, the stage in the growth of an insect during which time the body of the grub is broken down and reformed into an adult.

Resin A sticky substance made by many plants to close wounds. It also covers the leaf buds of some plants. Some bees and wasps collect resin and use it as a nest-building material.

Saliva The liquid produced in the mouth from glands in the head.

Social Living together in a colony.

Solitary Living alone — some bees and wasps do not live together in colonies but each female builds her own nest.

Thorax The second, or middle, part of an insect's body which bears six legs and two pairs of wings.

Finding Out More

If you would like to find out more about bees and wasps, you could read the following books:

Blau, Melinda E. *Killer Bees.* Milwaukee, WI: Raintree, 1983.

Boy Scouts of America. *Beekeeping.* Irving, TX: Boy Scouts of America.

Cook, David. *Small World of Bees and Wasps.* New York: Franklin Watts, 1981.

Dickson, Naida. *Biography of a Honeybee.* Minneapolis, MN: Lerner Publications, 1974.

Eastman, David. *I Can Read About Bees and Wasps.* Mahwah, NJ: Troll Associates, 1979.

Hawes, Judy. *Bees and Beelines.* New York: Harper & Row, 1972.

Powell, John. *The World of a Beehive.* Winchester, MA: Faber & Faber, 1979.

Reigot, Betty P. *Questions and Answers About Bees.* New York: Scholastic, Inc., 1983.

Index

Abdomen 10, 11, 21, 22, 24

Beekeepers 8, 15, 32
Bumblebees 8, 15, 32

Caterpillars 8, 17-19, 29
Cell 17, 20, 23-25, 28-30, 33, 34
Chitin 10
Cocoon 19, 23
Colonies 8, 28, 31, 33-35
Combs 30, 34

Cuckoo bees and wasps 8, 38, 39

Drones 34, 35

Eggs 8-10, 12, 19, 23, 25-27, 34, 35, 38, 39, 42
Enemies 38-41

Galls 26, 27
Glands 15, 35, 40

Grubs 8, 10, 12, 19, 23, 26, 28, 32, 35, 39, 40

Head 10
Hibernation 30, 31, 33
Hives 34, 35, 40, 43
Honey 32, 35, 39, 40, 42, 43
Honeybees 8, 14, 34, 36, 40, 43
Hornets 30

Ichneumon wasps 27

Leaf-cutter bees 24, 25, 44

Mason bees 22-24, 44
Mating 14, 15, 17, 29, 30, 33, 35, 42
Mining bees 20-22
Mud-dauber wasps 23

Nectar 11, 23, 25, 36, 42

Ovipositor 8-10, 27

Paper wasps *(Polistes)* 28, 30
Parasites 9, 26, 27

Pollen 11, 20, 23-25, 32, 33, 35, 36, 42

Queen bees and wasps 14, 28-35, 39
Queen substance 35

Resin 23
Royal jelly 35
Round Dance 37

Sand bees 21
Sawflies 8, 9
Skeleton 10

Swarm 35

Thorax 10
Thread-waisted wasps *(Ammophila)* 16-18, 20

Wagging dance 37
Wax 32, 34, 40
Wood wasps 27
Wool-hanger bees 15, 25
Worker bees 28, 32, 34-36, 39

Yellow jacket wasps 30, 31, 35, 40

Picture Acknowledgments

Prema Photos (R.A. Preston-Mafam) 21. All other photographs from Oxford Scientific Films by the following photographers: A. Bannister 41; G.I. Bernard cover, 8, 22, 37, 38, 39, 44; J.A.L. Cooke 9, 10 (top), 12 (bottom), 13, 27, 28, 30 (bottom), 42 (right); S. Dalton opp. title page; T.Owen Edmunds 40; S.tMorris 14, 16, 17, 19, 29; P. O'(bottom), 13, 27, 28, 30 (bottomToole 20; A. Ramage 15, 25 (right); D.M. Shale 32, 33; P.K. Sharpe 31; T. Shepherd 23, 24, 25 (left); D. Thompson 10 (bottom), 12 (top), 18, 26 (bottom), 34, 35, 42 (left), 43; G. Thurston 30 (top); P. & W. Ward 26 (top). Artwork by Wendy Meadway.